A Guide to PSILOCYBIN Mushrooms

by
RICHARD COOPER

Illustrations: **Graeme Jackson**
Alexandra King
Richard Cooper

HASSLE FREE PRESS

CONTENTS

Page

3	INTRODUCTION
4	HISTORY
5	IDENTIFICATION
9	PANAEOLINA FOENISECII
10	PANAEOLUS CAMPANULATUS
12	PSILOCYBE COPROPHILA
14	PSILOCYBE MONTANA
16	PSILOCYBE SEMILANCEATA
18	HYPHALOMA CYANESCENS
20	AMANITA MUSCARIA
22	CHEMISTRY AND DOSAGE
24	COLLECTING AND PRESERVING
26	EFECTS
27	SPORE PRINTING
28	SPORE DETAILS
29	LEGAL POSITION
30	GLOSSARY OF TERMS
31	GLOSSARY OF SCIENTIFIC NAMES
32	BIBLIOGRAPHY

A Guide to British Psilocybin Mushrooms. Text © 1977 Richard Cooper
Illustrations © 1977 Graeme Jackson.
All rights reserved. No part of this book may be transmitted, recorded or reproduced without prior permission of the copyright holders. This edition published 1994 by Hassle Free Press. First published by Red Shift Research Collective. ISBN 086166 0048. British Library Cataloguing in Publication Data. A catalogue record for this book is available from the British Library. Printed in the UK.

INTRODUCTION

This booklet is about the species of hallucinogenic mushrooms which grow in this country. Of the innumerable species of fungi native to the British Isles, about a dozen are psycho-active when taken in the correct dosage. *Psilocybe semilanceata*, the Liberty Cap, and *Amanita muscaria*, the Fly Agaric, are by far the most common. The purpose of this book is to help you find and identify these particular mushrooms, to give you some idea of what dosage to take, and to explain how to dry the mushrooms you find, for later use.

The mushroom which you see growing is not the whole fungus. It is only the part which produces the *spores*: the fruiting body or *sporophore*. The vegetative part of any mushroom—the part which absorbs nutrients from the soil and carries them to the fruiting body is a fine, thread-like structure which forms underground. These are called the *mycelium* of the fungus, individual threads being called *hyphae*. These networks can cover acres and one system may have more connections than a human brain.

Upon the germination of a spore in suitable conditions it puts down one or more fine germ tubes, which grow outwards from the spore. The tubes begin to branch and interlink, becoming more and more complex in their interlinking as the mycelial network forms. *Mycelium* from two spores must meet to enable a fruiting body to be formed. The first sign of development of a *sporophore* is the gathering together of the mycelium to form little knobs on the ground. In the Psylocibe a continuous ring shaped cavity forms below this precursor to the cap, into which the *hyphae* grow to form the radiating *gills*. This gill cavity is initially closed below by a membrane, the veil. As the mushroom grows this *partial veil* rips exposing the gills, portions of it remaining attached to the upper part of the stem and the *margin* of the cap, (this veil does not remain in *Psilocybe semilanceata*).

As the mushroom continues to grow and develop, spores are formed on the surfaces of the gills. When the spores are ripe they are shot off from these surfaces and drop downwards in the space between the gills. When clear, they are carried away by air currents to repeat the whole process. It is estimated that a 4" diameter mushroom will discharge 16,000,000,000 spores over a 6 day period.

HISTORY

Researching this pamphlet we have had to rely on mainly American historical sources, where published material on local hallucinogens is more advanced. We feel that there must be a history of Psilocybin use in Britain and Europe but have so far found no written material to draw from. We hope by writing this pamphlet to stimulate research into this neglected area of our essential folk history.

The first record we have found in modern times of the use of Psilocybin mushrooms comes from Spanish conquistadores at the coronation of Montezuma in 1502, who observed hallucinogenic mushrooms being served at the feast. Fray Bernadino de Sahagun (1547-1569), a Spanish friar, wrote about the Aztecs using the mushrooms as a sacrament under the name *Teonanacatl* or 'God's Flesh'. Sahagun says that the mushrooms "are harmful and intoxicate like wine. Those who indulge see visions, feel faintness of heart and are provoked to lust." Included in his text were drawings of some of the mushrooms.

In 1651, the king's physician, Hernandez, gave an account of the rituals in which the mushrooms were worshipped by Mexican natives. In this he says the mushrooms can "bring before the eyes all sorts of things such as wars and the likeness of demons". He also writes: "Yet others (mushrooms) there are not less desired by princes for their festivals and banquets, and these fetch a high price."

More recently, beginning in 1953, R.Gordon Wasson explored the area of Oaxaca in search of more information about the rituals and the mushrooms used. He and a friend became the first non-indians to attend a mushroom ceremony and ingest mushrooms.

In 1956 Wasson invited a French mycologist, Roger Heim, to research the use of sacred mushroom in the Oaxaca area. Heim identified fourteen species, three of which were unknown outside the area, but had been used by the indians for centuries. In 1958, Dr. Albert Hoffman, the first alchemist of LSD, isolated and named the two active agents as Psilocin and Psilocybin. Wasson also described frescoes depicting mushroom worship dating from 300 AD and mushroom stone images from Guatemala going back to perhaps 1000 BC.

It is safe to presume, though, that the hallucinogenic properties of these mushrooms and other naturally occuring plants were widely known in Europe for many centuries preceding the Industrial Revolution when much that was common knowledge amongst our rural ancestors fell into disuse. Probably the greatest single factor being the widespread prosecution of "witches" across Europe during the Middle Ages. Many of these so called "witches" were merely village herbalists whose knowledge of magical plants was a threat to the expanding power of the church. The witch trials were in fact a massive conspiracy to eliminate the last vestiges of a tradition of Earth religions, matriarchal in origin and incorporating magical plants,that stretched back to the beginnings of European civilisation.

IDENTIFICATION

These dozen species which can be hallucinogenic, may be divided into two distinct groups but the vast majority are all in one group. These are the mushrooms that can contain the drugs *psilocin* and *psilocybin*. The amount of drug present varies, depending upon the growing conditions. Some members of this family also contain quantities of the drugs *baeocystin* and *norbaeocystin*, which, like *psilocin* and *psilocybin*, are *alkaloid tryptamine* derivatives.

The second group contains the Amanitas which have *hyoscyamine, muscimol* and *ibotenic acid* as their psycho-active ingredients. It was thought that the poison *muscarine* was the main active ingredient in *A.muscaria* and *A.pantherina*, but recent research has shown that it is only present in minute quantities (0.00025% of the fresh mushroom) and has no clinical effect.

Group One: Psilocin and Psilocybin containing species

FAMILY	GENUS	SPECIES	Occurrence	Potency
	Psilocybe	Coprophila	O	+
		Montana	O/F	+
		Semilanceata	F/C	++
Geophilla	Stropharia	Aeruginosa	C	−
		Semiglobata	C	−
		Merdaria	O/F	−
	Hyphaloma	Cyanescens	O	+++
Coprinacae	Panaeolina	Foenisecii	F/C	+
	Panaeolus	Campanulatus	O/F	±
		Sphinctrinus	R/O	+

There are over 40 hallucinogenic species in the genus *Psilocybe*, three of which can be found quite easily in this country. *P.semilanceata* or the "Liberty Cap" is, by far, the most commonly occurring hallucinogenic mushroom. It is also the most reliable as it invariably contains both *psilocin* and *psilocybin* and can also contain *baeocystin*. *P.montana* and *P.coprophila* also grow readily in this country, but in nowhere near the profusion that *P.sem* occurs. Both contain the same psycho-active ingredients, but the drug content seems to be more variable, ranging from about the same as *P.sem* to virtually nothing. *P.coprophila* seems to lose its potency soon after picking.

Most of the other mushrooms of group one appear to display what is called a latent characteristic. That is, the drug content varies considerably according to the age of the specimen, its location and with the soil and weather conditions. Very little research has been done into the quantitative drug content of these mushrooms, and the necessary analysis involves the use of chemical extractions which are illegal to perform unless one is licensed to do so. Therefore much of the information in the following few paragraphs is based on what is known about these species when they occur in similar climatic zones.

The other mushrooms in this group, besides the psilocybes, belong to the genera *Panaeolus, Stropharia* and *Hyphaloma*. All four genera have similar characteristics and the distinctions which separate one genus from another are usually on the microscopic scale. Species which are listed here as belonging to a particular genus are often listed as being members of another genus inside this group. Notably *Panaeolina foenisecii* which is often listed as *Panaeolus foenisecii* or *Psilocybe foenisecii*.

In the genus *Panaeolus* three hallucinogenic species occur, all of which are latent. Specimens of both *Pan.campanulatus* and *Pana.foenisecii* have been shown to contain *psilocin* and *psilocybin* but always less than is present in *P.sem*. Other *Tryptamine* derivatives, particularly *serotonin* are often also present. *Panaeolus sphinctrinus*, the relatively uncommon sub-species of *Pan.camp,* which is distinguished from the latter by its permanently inturned margin, appears to be more consistently active.

Some specimens of *Pan.camp* collected in the northern United States contained *pantherine* and other toxins. If these toxins are present in British specimens the ingestion of any quantity would make you very ill.

In this country, *Stropharia aeruginosa*, a mushroom carrying a 3" flattened-convex, blue-green cap and *Stropharia semiglobata*, which is similar to *P.coprophila* save for its remnant veil and tawny cap, may be commonly found in the late summer and autumn. These, along with the less common *S.merdaria* are believed by many authorities to be active. At present I can find no conclusive evidence, but expect that they will only have a slight effect when found growing in this country. However, other members of this genus which are not native to Britain, contain ample quantities of psycho-active drugs. Notably, *S.cubensis*, which is common throughout the south-eastern United States, Central America and Southern Europe and contains about 0.2% (of dry wt.) of *psilocin + psilocybin*. This is the most commonly and probably the most easily cultivated psycho-active mushroom.

Finally in this group we have a solitary mushroom belonging to the genus *Hyphaloma*. *H.cyanescens* grows in wood-chips, straw and rank grass and has been found under cedar, elderberry, ivy and buddleia. This mushroom is similar to both *P.maztecorum* and the tropical *P.cyanescens* and in specimens collected in Algeria, the *psilocin* content was 0.5 - 0.6% of the dried weight of the mushroom, almost five times the concentration found in *P.sem*. *H.cyanescens* also contains sizeable quantities of *baeocystin* and *norbaeocystin*.

Group Two

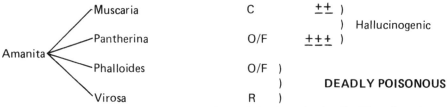

This group is composed solely of members of the Amanita family. These have been separated from the others, firstly because they contain different psycho-active ingredients and secondly because two species in this genus are deadly poisonous.

The two which are hallucinogenic when taken in the correct dosage are *Amanita muscaria*, the "Fly Agaric" and *A.pantherina*, the more potent "Panther cap". These contain *ibotenic acid*, *muscimol* and *muscazone* as their active ingredients, and have been praised for thousands of years for their magickal properties. All *muscarine* present

Gill structures

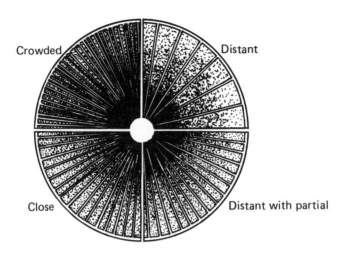

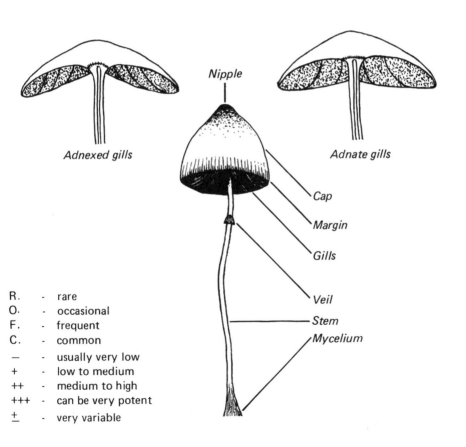

R. - rare
O. - occasional
F. - frequent
C. - common
— - usually very low
+ - low to medium
++ - medium to high
+++ - can be very potent
± - very variable

is in an inactive form. One problem which I feel I should point out to those of you who intend to experiment with these two mushrooms is the great variation in the quantities of the active ingredients which may be present. This can vary between 0.1 and 1.2%, by wt. of the dried mushroom. This combined with the relatively narrow margin between an effective and a lethal dosage make an overdose all too easy.

Fatalities from consumption of *A.muscaria* seem to be almost unheard of and in the one case of a fatality from *A.pantherina* that I have found recorded, something like 24 mushrooms were consumed! However lots of people get very ill by taking them.

Of all the deaths attributed to mushroom poisoning over 90% are caused by two members of this same family. *A. virosa*, the "Destroying Angel" and *A.phalloides*, the more common "Death Cap". These two speices are guaranteed to kill if eaten (50% will die). Therefore if you are going to eat wild mushrooms *A.virosa* and *A.phalloides* must be studied and recognised in all stages of their growth.

The deadly species have these specific characteristics:—

Spores:	White
Gills:	White, and free from the stem
Stem:	White, with the veil remaining
Base:	Bulbous, and growing out of a cup (remnants of the "Universal Veil")

A.phalloides, the most deadly, has a yellow to olive-green cap.
A.virosa is wholly white.

To the inexperienced eye entirely different species can appear to be very similar to one another; little pointy cream mushrooms look like little pointy cream mushrooms, but it is the subtler features, which are not seen at first, that distinguish one species from another. An hallucinogenic mushroom from an inactive one and an edible one from a poisonous one.

The genera most commonly mistaken for psilocybes are *Inocybe, Conocybe, Mycena* and *Bolbitus*. In the specific descriptions of each species, given later in this book, enough points of identification are given for absolute certainty. Check each point carefully. If they all tally, you have that mushroom.

There is a simple positive test for the presence of *psilocin*. If a mushroom is pulled up by the roots, the flesh of the stem, particularly that which was below the ground, will stain a blue-green colour on being bruised and exposed to the air for 2 - 3 hours. *P.sem* and *H.cyan* often have a blue-green base to their stems while still growing. This bluing reaction is due to the enzymic oxidation of psilocin.

N.B. *Inocybe Lacera* and *I. geophylla* superficially resemble *P. coprophila*. Both Inocybes contain *pilocarpine* which is a strong narcotic, acting in the same way as *muscarine*. Both Inocybes are woodland species whereas *P.cop* is a dung loving type. *I. geophylla* is lilac tinged and gives a brown spore-print. *I.lacera* has white gills and a grey-brown spore-print.

Without wanting to discourage you, I feel I should end this section on a note of caution. Previously unlisted species are regularly being found to be native in this country. Don't eat any fungus unless you know what it is and that it won't kill you.

PANAEOLINA FOENISECII (Haymakers mushroom)

Fruiting Season:		Grows from late spring, through the summer until early December and is most prolific in September and October.
Habitat:		Singly or in troops, in short grass, on lawns, fields, meadows and rich pasture lands.
Cap:	size:	1/3"-1" in diameter and 1/3"-2/3" high.
	shape:	Conic to hemispherical at first, becoming convex, with some specimens finally being umbrella shaped.
	colour:	Yellow-buff, and shiny, with light brown areas, when dry, becoming brown-purple when wet.
	texture:	Smooth with some lines when moist, cracked and scaly when dry, particularly older ones.
Stem:	size:	1"-3" high and up to 1/8" in diameter.
	shape:	Straight and even, occasionally slightly flattened.
	colour:	Pale tan with a red tinge when dry, becoming brown when wet. Always lighter than the cap, particularly at the top when young.
	texture:	Smooth, hollow and brittle with no veil remaining.
Gills:	colour:	White edged with brown faces, darkening and becoming mottled black with spores as the mushroom matures.
	attachment spacing:	Closely spaced, broad and adnately attached.

PANAEOLUS CAMPANULATUS

Fruiting Season:		April to December.
Habitat:		Solitary or grouped, on dung and rich well-manured soil, among long grass in pastures and meadows.
Cap:	size:	1/3"-1" in diameter, and slightly taller than broad.
	shape:	Parabolic to bell-shaped and never becoming flattened.
	colour:	Reddish-brown when wet, paler, almost white or grey and shining when dry.
	texture:	Wrinkled, and possibly cracked, when dry. Smooth and silky when wet. Thin-fleshed.
	margin:	Incurved at first, then expanding slightly. The partial veil remains as a white toothed fringe.
Stem:	size:	1½"-3½" in height, 1/16"-1/8" thick.
	shape:	Almost straight, of equal width, hollow, and slightly bulbous at the base with mycelium.
	colour:	Reddish at first becoming brown with age. The upper part is often spotted black with spores.
	texture:	Polished, slightly ribbed, and covered with tufts of minute white hairs which give the stem a mealy appearance.
Gills:	colour:	Grey-brown with a white edge becoming black mottled purple with age.
	attachment spacing:	Crowded with an adnate to adnexed attachment to the stem.

Panaeolus campanulatus

PSILOCYBE COPROPHILA

Fruiting Season:		This is the least common Psilocybe in this country and may be found from mid-August until mid-December.
Habitat:		Solitary to gregarious and almost invariably growing on dung.
Cap:	size:	¼" - 1" diameter and half as high.
	shape:	Hemispherical.
	colour:	Hazel to dark brown and shiny when wet.
	texture:	Viscid or dry, with a slight roughness around the margin, particularly when young.
	margin:	A toothed white fringe, remnants of the partial veil, stays until late maturity.
Stem:	size:	1" - 3" high and about 1/8" thick.
	shape:	Usually quite straight and parallel sided, with a base slightly enlarged by the mycelium.
	colour:	Yellow brown and slightly paler than the cap.
	texture:	Dry, smooth, stiff and cartilaginous with little or no remnant of the veil.
Gills:	colour:	White edged with grey-brown faces which become purple or almost black with age.
	attachment spacing:	Broadly spaced almost triangular gills which are adnate and squarely attached.

Psilocybe coprophila

PSILOCYBE MONTANA

Fruiting Season:		From late July until early December.
Habitat:		Among mosses, lichens and ferns, in sandy soils in woods.
Cap:	**size:**	¼"-1" in diameter and about half as high as wide.
	shape:	Hemispherical to slightly expanded and distinctly breast-like, often with a pronounced nipple.
	colour:	Tawny-brown when wet, grey-brown when dry.
	texture:	Smooth, thin-fleshed and ridged to half way up the cap.
Stem:	**size:**	1"-3" high and about 1/8" thick.
	shape:	Usually of even thickness and slightly wavy with a flare at the base where the mycelium joins the stem.
	colour:	Brown, but slightly paler towards the cap.
	texture:	Smooth and without remnants of the veil.
Gills:	**colour:**	Light brown when young becoming dark brown with age.
	attachment **spacing:**	Broadly spaced, adnate and almost triangular.

Psilocybe montana

PSILOCYBE SEMILANCEATA (Liberty Caps)

Fruiting Season:		This is the most common species of Psilocybe, and can be found from late August to mid-January, most of the fruits appearing in September and November.
Habitat:		Amongst grass, in fields, heaths, parks, meadows and roadside verges. They are frequently found in large troops, and like a well manured soil, but do not grow in dung. If the soil has been kept wet throughout the heat of the summer, (ie. lawns, football and cricket pitches, etc.,) far more mushrooms can be expected.
Cap:	**size:**	3/16" - 1" in diameter and about 1½ times as high.
	shape:	Hemispherical to conical, with a prominent nipple on the top. The margin of the cap is usually incurved.
	colour:	Buff when dry, becoming yellow-brown with a marked olive tinge when wet.
	texture:	Smooth, thin fleshed and pliant when wet. Thin black lines can be seen around the margin, particularly in older specimens, where the gills show through the flesh of the cap.
Stem:	**size:**	1" - 3" long and up to 1/10" thick.
	shape:	Wavy and equal with a slight flare towards the base where it connects with the mycelium.
	colour:	White to cream and paler than the cap. Very often there is a slight bluing at the base of the stem.
	texture:	Smooth, tough and pliable with no remnants of the partial veil.
Gills:	**colour:**	The edges of the gills are white, their faces being cream coloured when young becoming dark brown-black, tinged purple with age.
	attachment spacing:	Adnate to adnexed and crowded.

Freshly picked *P. sem.* in dry conditions.

Freshly picked *P. sem.* in wet conditions showing stages of growth.

Psilocybe semilanceata (Liberty Cap)

All Photographs by Chris Render

Small crop of *P. sem.* 30 mins after picking.

The same crop of *P. sem.* after 24 hours

Psilocybe semilanceata

HYPHALOMA CYANESCENS

Fruiting Season:		August to late December
Habitat:		Singly or in small groups, on decayed wood, on tree stumps and in humid grass.
Cap:	size:	1"-2" in diameter and about half as high.
	shape:	Conical when very young becoming breast-like and finally plane or even uplifted and wavy.
	colour:	Yellow when dry and chestnut-brown when wet, staining blue when touched.
	texture:	Smooth and firm when dry and slightly sticky when wet. More fragile with age.
	margin:	After the cap is flat the margin continues to grow and so buckles and becomes wavy.
Stem:	size:	2"-4" long, 1/8"-1/4" thick.
	shape:	Equal and slightly bent with a noticeable flare for the bottom inch or so. Some signs of the veil are usually present.
	colour:	White at the top, amber towards the base, going blue when dry.
	texture:	Brittle with silky fibres on the surface.
Gills:	colour:	Cinnamon becoming reddish-brown with black-purple splotches as the spores mature.
	attachment spacing:	Distantly spaced and adnately attached with partial gills in between.

Hyphaloma cyanescens

AMANITA MUSCARIA (Fly Agaric)

Fruiting Season:		Very common between late July, if a wet summer, and early December. Most common in September
Habitat:		Solitary and in scattered groups beneath birch, beech and hazel, pine and spruce trees. Usually on poor soils.
Cap:	size:	1½"-8" in diameter and 1½"-2½" deep.
	shape:	Almost spherical when young becoming convex to umbrella, and then plane to concave with age.
	colour:	Scarlet-red to orange when young then more orange-yellow as it expands and ages. Normally speckled with small white spots 1/16"-½" in diameter.
	texture:	Tough and fleshy. The surface is usually covered with white warty spots, remnants of the universal veil, which can quite easily be washed off by a heavy rainfall.
	margin:	Striate upward for ¼"-¾".
Stem:	size:	3"-10" tall and ¾"-1½" in diameter.
	shape:	straight and parallel, usually expanding gradually to a swollen base.
	colour:	Shiny white with yellow highlighting.
	texture:	Ragged and scaly, particularly towards the base, where fragments of the universal veil remain in broken rings. Viscid in wet weather. The partial veil remains as a hanging skirt towards the top of the stem.
Gills:	colour:	White with a creamy edge.
	attachment spacing:	Crowded, narrow and freely attached.

AMANITA PANTHERINA

This mushroom is less common but slightly more potent than *A.musc.* and can be easily distinguished from the latter by the following characteristics.

It is slightly smaller, a specimen with a 5" diameter cap being relatively uncommon. The cap is dark brown when young, speckled white with remnants of the Universal veil, becoming coffee coloured with age. The stem is tan and not white, the scales at its base forming complete concentric circles rather than broken rings.

Amanita muscaria

CHEMISTRY AND DOSAGE

The main active drugs found in the group one mushrooms are the *tryptamine* derivatives *psilocin*, *psilocybin* and *baeocystin*, all of which contain the *indol* group characteristic of hallucinogens. The *tryptamine* molecule is composed of this indol group and a *dimethylamine* chain, $-CH_2.CH_2.NH_2$

In *psilocin* and *psilocybin* the two hydrogen atoms at the end of the dimethylamine chain are replaced by methyl groups $-CH_3$. In *baeocystin* only one hydrogen atom is substituted in this way. The length of chain in these substituted groups affects the intensity and duration of the drugs action.

In *psilocybin* and *baeocystin* one hydrogen in the benzene ring (4 position) is replaced by a *phosphorloxy* group $O.PH_2.O_3$. On ingestion of *psilocybin* the phosphorloxy group is hydrolised to an hydroxy group, $-OH$, giving *psilocin*.

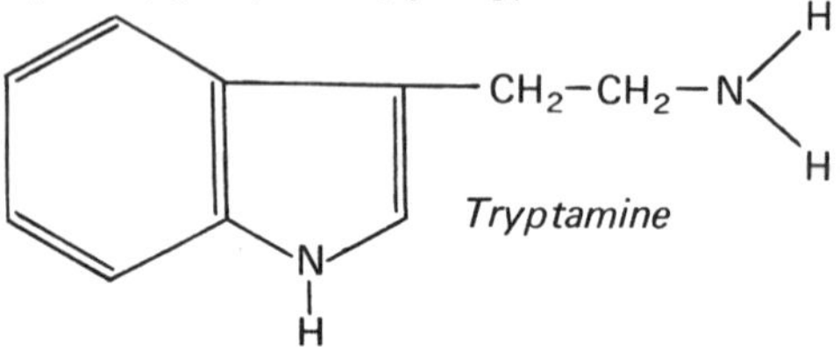

Tryptamine

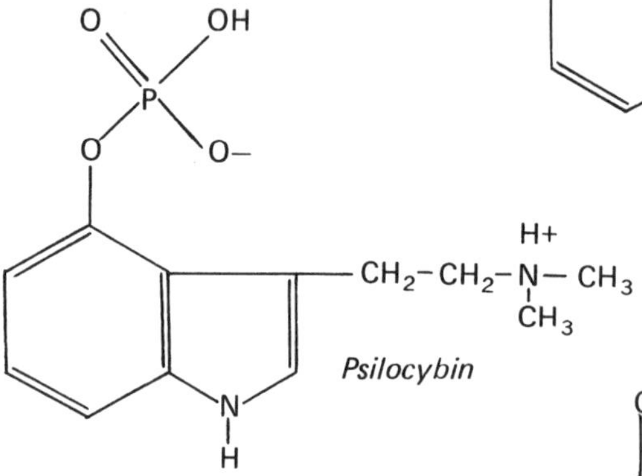

Psilocybin

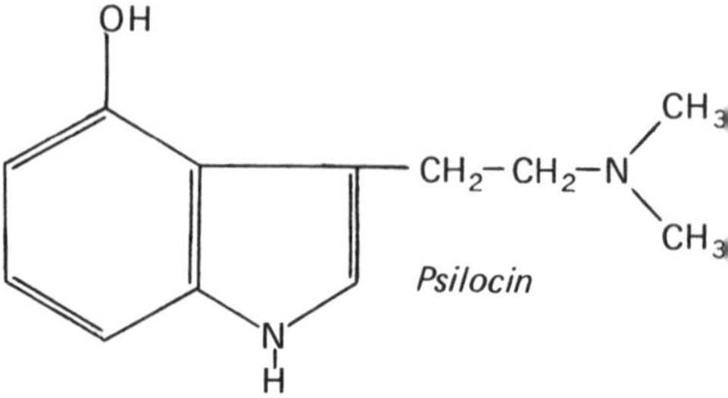

Psilocin

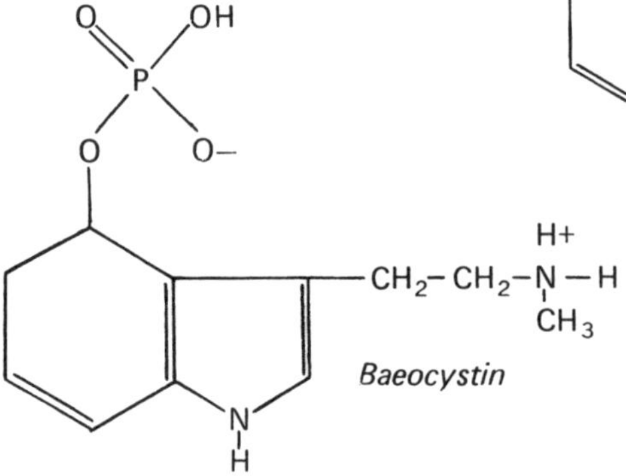

Baeocystin

It takes about 50% more *psilocybin* than *psilocin* to produce the same effects, and at present no information as to the specific actions of *baeocystin* and *norbaeocystin* is available.

All these drugs are thought to produce their psycho-activity by altering the serotonin (another indol) levels in the brain thus interfering with the transmission of stimuli that regulate the processing of information. A dosage of 3-6mg of *psilocin* (0.1mg/kg) produces hallucinations in man and doses as low as 0.5mg are discernible.

As the dosage is increased beyond 6mg the effects are noticed much sooner, seem to be more intense and last slightly longer. The lethal oral dose of *psilocin* in man (50% will die) is estimated to be 100mg/kg or about 80lbs. of fresh *p.sem*. This LD_{50} figure does not mean that a massive dose, below this figure, will not cause irreparable damage.

The group two members, *A.musc* and *A.panth*, both contain *ibotenic acid* and its derivative *muscimol* as their main psycho-active ingredients. These *oxazole* derivatives affect both the central nervous system and the sympathetic nervous system, acting as sedatives as well as hallucinogenics and producing both physiological and psychotomimetic effects. Orally the potency of *ibotenic acid* is one tenth that of *muscimol*. The effective doses, in man, being 70mg and 8.5mg respectively. The lethal doses, in man, of *ibotenic acid* is 40mg/kg and of *muscimol* 45mg/kg.

Ibotenic acid

Muscimol

As previously mentioned the quantities of psycho-active drugs present differ from specimen to specimen of a particular species and this quantity is dependent on the soil and climatic conditions as well as the size and age at the time of picking. With so many variables no hard and fast rules on dosage can be formulated in terms of the numbers of mushrooms to eat to achieve a particular effect. One thing which must be pointed out about oral consumption of drugs is the fact that drugs taken on an empty stomach have a far stronger effect than when ingested on a full stomach. Taking 10 *P.sem* on an empty stomach can be like taking 40 after a meal and when consumed with alcohol they are even stronger.

Of group one, *P.sem* can be considered the most consistent in its content and be used as a "yard-stick" by which to assess the others.

It contains between 0.1% and 0.4% *psilocybin* by wt. of the dried mushrooms and up to 0.2% *psilocin*. From this one can deduce that 30-40 gms of fresh *P.sem* or about 5gms of dried *P.sem*, that's about 30 small mushrooms, can be expected to contain 3-12mg of *psilocybin*. This figure is usually about 6mgs, an effective dose.

P.cop, *P.mont*, *Pan.camp*. and *Pana.foen*. can all be expected to contain smaller quantities of these drugs while *H.cyan* may contain up to five times the amount given above.

The active drug content of *Amanita muscaria* and *Amanita pantherina* varies from virtually nothing to as much as 1.2% of the dry weight. This content is governed by the same factors which affect psilocybes. 1gm of dry *A. musc* can be expected to contain 1mg of *ibotenic acid* and 3mg of *muscimol*, but these figures can be as high as 5mg and 10mg respectively. *A.panth* is usually considered to be stronger than *A.musc*. So although 4gms of dried mushrooms will normally constitute a reasonable dose these 4gms may contain 6 times the amount you expected.

If you intend to take psychotropic mushrooms be careful, particularly with the Amanitas. Take a small dosage to check the level of the drug content before you contemplate taking a full dose.

A good technique to assess the quality of a particular batch is to do 3 controlled experiments, each on a different occasion. It is better if these experiments can be carried out with at lease 4 days between each as a tolerance to the drug may affect your results. In the first experiment eat a very small amount, maybe 5 psilocybes or ¾gm of dried Amanita or 1½ *H.cyan* and try to remember what it feels like. In the second experiment take twice the amount. Again try to assess the effect, but this time contrast it with the experience of the first experiment. For the third test, double the dosage again (that is if nothing untoward has happened so far) so that you will now be taking about 20 psilocybes or 3gms of Amanita or about 8 H.cyan. By now you should have a good idea of the comparative effect of different doses and be able to assess for yourself what constitutes a "normal" dosage.

Please take care.

COLLECTING AND PRESERVING

The main "flush" of psycho-active mushrooms occurs between early September and late November but they can be found as early as the first week in August. A preliminary expedition in mid-August can be very useful to determine where large areas of mushrooms might grow later in the season.

When you go on a hunt for magic mushrooms it's worth remembering that you are more likely to find active specimens of *Psilocybe semilanceata* than active mushrooms of any other species. As *P.coprophila*, *Pan.campanulatus* and *Pana. foenisecii* grow in much the same conditions as *P.sem.*, it's worth heading for a rich grassy area first. In cities, football pitches, cricket squares, lawns in parks and formal gardens are very likely spots. They are well manured and usually kept well watered throughout the summer, this keeps the soil rich and moist, giving the mycelium more of a chance to grow throughout the summer. Places which retain moisture naturally are also good locations; fields near rivers or lakes, low ground etc., mushrooms also favour areas where the ground is warm, so look for an area where sunlight will fall for most of the day.

If, after wandering around at the would-be picking site, no areas of fungi are noticeable, a more detailed search is called for. The Psilocybes are quite small and to see to find them your eyes will need to be at the most 4 feet from the ground. Search quite a large patch to start with, maybe 40 ft. square. If you don't find anything cover another similarly sized area. If you still don't find anything look closer. If you still have no joy, try searching about 60-70 yards away from where you started. Still no luck? Get down closer, separate long grass with your fingers and look underneath. Don't be too discouraged if you don't find what you're looking for, it may be too early in the season, or not a particularly good site. Try a different site, or a different day, or both.

When you do find a mushroom don't wander away from that area as there could be others, close by, growing off the same mycelial network. Check its visible characteristics against those given. If they all tally you have a psilocybe. One marked characteristic of psilocybes is that the mycelium, where it joins the stem, often has a blue-green tinge. This bluing also occurs within 10-15 minutes if you scratch or bruise the mushroom's flesh.

Pick them by twisting the stem sharply just above the point where it reaches the ground, or nip the stem with your fingernails. If you pull them up by the roots this will damage the surrounding mycelium and slow down the production of further fruiting bodies. Don't pick all the little ones as they'll be big ones tomorrow.

It's best to be sure what you're eating before you consume any wild mushrooms. When you get home take a spore print (see cultivation). If it's not purple/black—black-purple/brown you have a different genus which might be poisonous.

When harvested the mushrooms can either be eaten fresh, when they have a very pleasant mild flavour, cooked with food, served as a tea or dried for later use (see legal position). *P.sem* loses 20-40% of its potency on drying. *P.cop* loses up to 80%

Traditionally mushrooms are collected in shallow baskets made of wood slats but a seed tray, a shallow cardboard box or a frisbee do just as well. If not too wet and fragile they can be transported in an unsealed plastic bag, otherwise move them in the collecting tray. Take care not to bruise the flesh as you handle them and do not overfill your tray or bag as those at the bottom will get squashed by the weight of those above.

When looking for small woodland species, like *P.montana* or *H.cyanescens* turn over fallen leaves, twigs and other debris, close to the base of trees and look underneath. *A.muscaria* sticks out like a sore thumb, but *A. pantherina* is much better disguised. Both these are often found in the birch and hazel hedges which are used to fence off country roads bordered by ditches.

Air and oven drying both work well, the potency remaining slightly greater when air dried. To air dry your mushrooms lay them out on sheets of newspaper so that they are not touching one another. Alternatively, if you don't have very much space, they can be strung together like beads by sewing them through the point of the cap with a needle and thread, making sure to leave a small gap between each to prevent them sticking one to another. Leave them in a warm dry place until they are dry. This drying will take between 6-48 hours depending on your drying conditions.

Do not discard the stems as they are equally potent

When the mushrooms are completely dry, hard and no longer spongy to the touch, they can be stored in sealed plastic bags. Mark on each bag the number it contains as the mushrooms are very crumbly in the dried state.

To oven dry your mushrooms spread them out on baking trays and place them in a low oven (below 150°F) until dry. If you put wet mushrooms straight into the oven they tend to stick to the trays. Although it takes longer, I find air drying more satisfactory.

NEVER KEEP UNDRIED MUSHROOMS IN A PLASTIC BAG FOR TOO LONG AS THEY TURN INTO A STICKY MESS.

EFFECTS

The experience you will have "on" hallucinogenic mushrooms is affected by a huge number of parameters, the most influential being; you yourself, your environment, and your companions during the trip. Your state of mind at the time; elation or depression, calmness or anxiety, satisfaction or frustration, will be of utmost importance, as it is chiefly that which will determine what you experience. Trips taken indoors can be very soul-searching and introverted, while those taken outside, especially in the countryside, seem to revolve around nature itself. The essential beauty and perfection of creation being overwhelmingly emphasized. Try to be with people you love and trust as this can go a long way in helping you to relax into what is happening. Regardless of whether or not you enjoy the trip you will learn something about yourself and your relationship with everything around you. If you don't want to experience too much mental discomfort it would probably be better to take these mushrooms when you are in a reasonably balanced frame of mind. If the trip does get too uncomfortable try to remember that what you are feeling is being greatly accentuated by a drug the effects of which will wear off in quite a short time. If that doesn't help, try doing something different, go for a walk, listen to some music, take a bath, anything that may change the stimuli that are causing your discomfort.

You will notice the initial effect of Psilocybin more quickly than you do with LSD (disregarding the initial high many people experience at the moment of ingestion of any drug). The state of mind induced by a full dose is one of calm euphoria, with no loss of coherence or clarity of thought. The initial effects of a moderate dose, 6-12 mg. can be felt within about 20 minutes. These first tell-tale signs are usually; mild euphoria, an apparent brightening of colours, flickering or shifting of the visual field and unusual auditory perceptions. The trip will get stronger and stronger for about 3 hours by which time you may be experiencing: strong auditory and visual hallucinations, particularly with your eyes closed, total euphoria and an overwhelming sense of belonging to a greater whole. After another hour or so the effects will gradually begin to subside. The total time lapse between ingestion and reasonable normality will be 6-9 hours which can be compared with 8-15 hours for a similar dosage of LSD (125 micrograms).

A tolerance to Psilocybin can develop quite quickly if taken more frequently than once every week to 10 days (there is also a cross tolerance with other hallucinogens). To achieve the same effect the next day, 1½-2 times as many mushrooms are needed. So when a tolerance develops you can either increase the dose or stop taking them until your body's metabolism has had time to regain its normal equilibrium. The latter course seems sensible, although, since the toxicity is so low, raising the dosage for a while is a viable alternative.

Little hard evidence is available as to what happens with prolonged continuous usage of large doses but long term mental distortions may well occur. As with all drugs, respect and caution are necessary.

When you wake up the next day, after taking a normal dosage, you will usually feel completely refreshed by your experience. Slight effects or memories of effects, may occur for up to four or five days and sometimes many weeks later.

SPORE PRINTING AND CULTIVATION

Cultivation of a pure strain of Psilocybin mushrooms is be no means a simple process as operations which require very sterile conditions and great patience are involved. However a brief outline of the process is given below.

The first step is to collect the spores of a known species of psilocybe. This is done by taking a spore print. To do this, take a sharp knife and cut the stem off a mushroom as close to the gills as possible. Place the mushroom, gills down, on a sheet of white glazed paper (it can be quite helpful to see the colour of the spores against different colour backgrounds). Cover the cap with a small bowl, to protect it from draughts or accidental movement, and leave it for 24 hours. Upon careful removal of the bowl and cap you should be left with a symmetrical deposit of spores. If the spores are to be examined microscopically it is a good idea to take the print on glass.

FROM THIS POINT UNTIL CASING ALL PROCESSES MUST BE PERFORMED IN ABSOLUTELY STERILE CONDITIONS.

For a mushroom to form, *mycelium* from two different *spores* must come into contact. This is achieved by germinating the spores on a nutrient medium, such as malt extract agar. If each culture dish is inoculated with spores in 3 or 4 places when the mycelium growing from two of these points meet, it will be from two different spores. Generally on a germination plate there are at least a dozen different strains of the same type of mycelium growing. These separate strains need to be isolated, one from another, to determine which grow and fruit most vigorously. This is done by inoculating fresh culture plates, in one place only, with small areas of the spore germination plate. It's a good idea to do at least a dozen plates as contamination may occur in some. The smaller the area you take from the culture plate, the greater your chance of isolating a pure strain. If, after a few days growth, a small area of each of these plates is again inoculated onto one of a fresh group of culture plates, the resulting strains will almost certainly be pure. Some strains will be seen to grow more quickly than others, spreading to cover the plate in less time than the slower ones. Three or four of those which exhibit strong mycelial growth should be kept and put through several fruiting stages to determine which one will fruit most vigorously.

Having decided which strain you wish to use the next stage is to get the volume of this strain to increase. This is done by inoculating jars full of sterilized grain and nutrient with the chosen mycelium. This is allowed to grow in warm, humid, sterile conditions (80°F, 95% humidity) for about 4 days, by which time most of the surface of the grain will be covered. This surface growth is then mixed throughout the grain by shaking and left for about 8 more days, by which time it will have completely permeated the contents of the jar. It is in this stage, of growth on grain, that contamination is most likely to occur and absolute sterility at all stages is the only way to avoid this.

The next, and final, stage in mushroom production is called casing. First you must cover the surface of the mycelium in the jars with ½"-¾" of moist, sterilized soil. These jars must then be kept warm and moist, in a humid atmosphere, for about 3 weeks during which time the mycelial network will penetrate the casing soil and begin to form into mushroom "primordia" (baby mushrooms). These will mature into *sporophores* within 10 days and will continue to mature for a further week or so. They will then disappear for a period of about 10 days after which another "flush" of mushrooms will appear. The harvest should last about 3 months in all, the final flushes yielding less and less fruits

If you are going to try your hand at cultivating psilocybes I suggest that you consult a good book on mushroom cultivation and one that deals with inoculation techniques in particular. Although it deals specifically with *P.cubensis*, Oss and Oerics's book (see bibliography) is to be highly recommended, the cultivation process being similar for most species.

Mushrooms can also be cultivated by the careful transplantation of sections of mycelium containing turf, from a natural growing site to a preferred site of similar soil character. These turfs should be about 1 sq.ft. in area and at least 4" thick.

SPORE DETAILS

Panaeolina foenisecii	12-17 x 7-9 microns, warty and elliptical with a pore at the apex giving a dark purple-brown spore print.
Panaeolus campanulatus	13-18 x 8-13 microns, elliptical with a pore at the apex giving a brown to purple-black print.
Psilocybe coprophila	11-15 x 7-9 microns, smooth and elliptical with a pore at the apex, giving a dark brown to purple print.
Psilocybe montana	6-8 x 4-5 microns and elliptical giving an umber-violet print.
Psilocybe semilanceata	12 - 14 x 7 - 8 microns and lemon shaped giving a purple brown print.
Hyphaloma cyanescens	9-12 x 7-9 microns smooth, oblong to ovate giving a purple black print.
Amanita muscaria	9-11 x 6-8 microns, smooth and ellipsoid giving a white spore print.
Amanita pantherina	10-15 x 5-8 microns, smooth and elliptical giving a white spore print.

LEGAL POSITION

Much confusion has surrounded the laws concerning the possession of hallucinogenic mushrooms, particularly since the late 1960s when they started to become increasingly popular as a mind altering drug.

Many people have been convicted for possession of mushrooms although convictions often occur when the police find them in the possession of someone they suspect of separate, unrelated drug offences and when no concrete evidence is available to secure a conviction for the suspected offence.

This situation was accidentally sorted out by the Law Lords in Sept. 1977 when the drug laws were changed to include the leaves and roots of *Cannabis Sativa* as controlled substances, along with the already controlled fruiting and flowering tops of this plant. This change in the law and various subsequent court cases have set up a very strong precedent that no drug substance is illegal unless it is specifically mentioned in the Misuse of Drugs Act. This legal point was eloquently highlighted by Judge P.M. Blomefield during his summing up at the end of a mushroom case at Reading Crown Court in April 1976, when he drew the jury's attention to the following points;

"In section 5(2) of the Misuse of Drugs Act 1971, Psilocin and Psilocybin are listed together with a large number of substances, some of natural origin, most of synthetic origin, as controlled drugs".

"Therefore the question before us is whether or not the possession of the Psilocybin mushroom, from the which the drug is extracted, and the fact that it contains the drug, amount to an offence."

"In this section, two natural vegetable substances are mentioned, coca-leaf and poppy straw. Also listed are the products of these substances, *cocaine*, *opium*, *morphine* etc. In these instances Parliament thought it necessary to specify the vegetable matter as well as the derived drug, suggesting that if you want to make possession of a substance, albeit a natural vegetable substance, an offence, it must be included specifically in the schedule."

Judge Blomefield also pointed out that Morning Glory seeds, which contain LSD— a substance listed in Schedule 2 and some species of toads (*bufos*) which contain a Class A drug of Scedule 2 are not listed either. He concluded therefore, that as Psilocybin mushrooms are not listed it is not an offence to possess them.

This all seems very rosy. It is. But beware! Don't forget that it is still an offence to make a preparation containing Psilocin or Psilocybin and mushroom soup, omelettes, tea or wine could all be construed by the authorities to be preparations. However it is very doubtful that possession of dried mushrooms can be considered an offence as they could have been picked after being dried by the sun.

GLOSSARY OF TERMS

Adnate:	With the entire width of the gills attached to the stem.
Adnexed:	With the gills narrowly attached to the stem.
Family:	A group including all related genera.
Genus (pl. genera):	A group of related spieces sharing certain similar characteristics.
Fungus (pl. fungi):	A non-flowering plant, devoid of chlorophyll.
Hypha (pl. hyphae):	One or more filamentous fungal cells.
Inoculate:	To implant micro-organisms into a culture medium.
LSD:	Lysergic Acid Diethylamide (an hallucinogen).
Micron (abbrev. μ):	Unit of metric measurement used for microscopic work. 1000 μ = 1 millimeter.
MUODA:	Misuse of Drugs Act.
Mycelium:	The mass of threads (hyphae) from which the mushrooms grow.
Mycophile:	One who likes mushrooms.
Mycophobe:	One who strongly dislikes mushrooms.
Partial Veil:	A membrane extending from the margin of the cap to the stem.
Psychoactive: Psychotomimetic:	Capable of altering the functions of the nervous system.
Species:	One or more individuals which show similar characteristics passed on from generation to generation.
Spore (pl. spores):	reproductive bodies of fungi.
Sporophore:	A mushroom.
Trip:	A term used to describe effect experienced upon ingestion of an hallucinogen.
Type:	A variety of the normal species.
Universal Veil:	A membrane surrounding the developing mushroom button (*Amanita*)
Viscid:	Moist and sticky.

GLOSSARY OF SCIENTIFIC NAMES

ABBREVIATION	GENERIC NAMES
A.	Amanita: Amanos, a mountain in Asia Minor which abounded in edible fungi. (c. 1000 BC).
H.	Hyphaloma: Hyphe—a webbed, loma-fringe.
I.	Inocybe: Inos—fibrous, kube—head.
M.	Mycena: Mukes - fungus.
Pan.	Panaeolus: Panaiolos—all variegated, which refers to the mottled gills.
Pana.	Panaeolina: Deminalive of panaeolus.
P.	Psilocybe: Psilos—bald, kube—head.
S.	Stropharia: Strophos—a sword belt, referring to the remnant veil.

SPECIFIC NAMES

aer.	aeruginosa: aeruginosus—covered with verdigris.
camp.	campanulatus: campanula—a little bell, -us—-shaped.
cop.	coprophila: copro—dung, philos—liking.
cub.*	cubensis: growing in Cuba.
cyan.	cyanescens: cyan—a green-blue colour.
foen.	foenisecii: foenisecia—the hay harvest.
mont.*	montana: growing in the mountains.
musc.	muscaria: muscarius—relating to flies, referring to A.musc.'s use as a fly poison.
panth.	pantherina: pantherinus—spotted like a panther.
phal.	phalloides: phallos—phallus.
sem.	semilanceata: semi—half, lanceatus —spear-shaped.
semig.	semiglobata: semi—half, globatus—spherical.
sphinc.	sphinctrinus: sphincter—a band, (holding in the margin of the cap).
vir.	virosa: viros—poisonous.

* This naming refers to where they were originally found: cubensis doesn't only grow in Cuba and montana grows in non-mountainous areas.

BIBLIOGRAPHY

Carter, Michael: *Will the legal Liberty Cap cause Home Office hallucinations?* New Scientist, Vol. 71 no. 1018 16 Sept. 1976.

Duffy, T.J. and Vergeer, P.P.: CALIFORNIA TOXIC FUNGI. 1977. Mycological Society of California Inc.

Emboden, William: NARCOTIC PLANTS, 1972. Studio Vista, London.

Furst, Peter T.: FLESH OF THE GODS, *the ritual use of hallucinogens*. 1972. Allen & Unwin, London.

Haard, Richard and Karen: POISONOUS AND HALLUCINOGENIC MUSHROOMS 1975. Cloudburst Press.

Lange, M. and Hora, F.B.: GUIDE TO MUSHROOMS AND TOADSTOOLS. 1963, Collins, London.

Menser, Gary: HALLUCINOGENIC AND POISONOUS MUSHROOMS 1977. And/Or Press, Berkeley, California.

Norland, Richard: WHAT'S IN A MUSHROOM, Part 3, 1976. Pear Tree Publications, Oregon.

Oss, O.T., and Oeric, O.N.: PSILOCYBIN MAGIC MUSHROOM GROWERS GUIDE. 1976, And/Or Press, Berkeley, California.

Ott, Jonathan,: HALLUCINOGENIC PLANTS OF NORTH AMERICA 1976. Wingbow Press, Seattle.

Schultes, R.E. and Hoffman, Albert: THE BOTANY AND CHEMISTRY OF HALLUCINOGENS. 1973. Charles C. Thomas, Springfield, Illinois.

Smith, Michael V.: PSYCHEDELIC CHEMISTRY. 1973. Rip Off Press, San Francisco.

TRANSACTIONS OF THE BRITISH MYCOLOGICAL SOCIETY SUPPLEMENT, 1960.

Wakefield, E.M. and Dennis, R.W.G.: COMMON BRITISH FUNGI. P.R.Gawthorn Ltd., London.

Wasson,R.: SOMA, DIVINE MUSHROOM OF IMMORTALITY, 1968, Harcourt Brace.

Villee,Claude A.: BIOLOGY, 1967. W.B. Saunders, Philadelphia and London.